PERPETUAL
ENVIRONMENTAL FORCES

PERPETUAL ENVIRONMENTAL FORCES

(Original Scientific Theory)

An Original Scientific Theory Setting Forth an Explanation of the Formulation of All Matter, Plant and Animal Life and the Mutation Changes that Occur Starting With the "Big Bang" and up to the Present and into the Future. Most Importantly, This Presents an Explanation of How Perpetual Environmental Forces Cause Mutations to Occur.

The Answer To Why We DREAM?

Sexual Orgasm—An Adaptable Characteristic

SYLVIA GOLDSTEIN

authorHOUSE®

AuthorHouse™
1663 Liberty Drive
Bloomington, IN 47403
www.authorhouse.com
Phone: 1-800-839-8640

Published by AuthorHouse   06/03/2013

ISBN: 978-1-4817-3803-3 (sc)
ISBN: 978-1-4817-3804-0 (e)

Library of Congress Control Number: 2013906627

Contents

Dedication

This book PERPETUAL ENVIRONMENTAL FORCES (Original Theory) is dedicated to my loving husband David. With his support and patience made this book possible. I thank him profusely. David—husband, most unusual. He started from poverty—high school dropout to become a successful attorney at law.

I further dedicate this book to my three loving sons. Richard Elliot Kaiser, Mitchell Lloyd Kaiser, and Gary Mark Kaiser who encouraged me.

I also dedicate this book to Richard Kaiser Jr., Matthew Kaiser, Jessica Kaiser, Adam Kaiser, Helane Kaiser, Phillip Kaiser, Tracey Kaiser, Haley Kaiser, all of whom, I deeply love.

In addition, I dedicate this book to my twin sister, Sandi Lehman, Larry Balish and Linda Zimmern, David Zimmern and Constance Levy Kaplan who are most helpful and supportive.

I further dedicate my book to Karen Gansel, Richard Gansel, Leslie Gansel, Susan Markowitz, Martin Markowitz, Blair Markowitz, Harris Markowitz, all of whom I greatly admire.

Preface

There is an age-old question—WHERE AND HOW DID ALL LIFE START?

I believe that I have found a scientific answer to man's quest to learn how all life started and the changes that occur and how man appeared on Earth (as well as all matter, plant and animal life) which include all cells, from the beginning leading to the present and into the future.

This answer is to be found in my original theory, which I call, PERPETUAL ENVIRONMENTAL FORCES.

This theory deals with perpetual environmental forces such as gravity, electromagnetism, sun heat, platetechtonics, etc., which forces, scientists have acknowledged exist.

What follows herein is an explanation of how the perpetual environmental forces answer the age-old question of WHERE AND HOW DID ALL LIFE START? and most importantly, how the changes occur.

THE ANSWER TO WHY WE DREAM?

SEXUAL ORGASM—AN ADAPTABLE CHARACTERISTIC.

Explanation and Observations

Presently we know there are mutations. A mutation is a genetic change to the reproductive cells of all living animal and plant matter. At this point in time, there has not been an answer to what causes a mutation. My original theory that I have entitled herein, PERPETUAL ENVIRONMENTAL FORCES, gives an answer to what causes a mutation. I came to this conclusion in the year 1984. Mutations give rise to adaptable characteristics enabling survival.

In the scientific world, it is accepted that all matter, plant and animal life evolves. However, what has not been set forth to date, is a scientific explanation or theory describing how such matter, plant and animal life, came into existence and further developed. After reviewing Sir Isaac Newton's theory of gravity, Charles Darwin's theory of evolution and Barbara McClintock's, (1983 Nobel Prize winner in the field of physiology) theory of transposable elements, I have come to the conclusion and theory regarding the formation of all matter, plant and animal life, which theory I call PERPETUAL ENVIRONMENTAL FORCES. Perpetual Environmental Forces is an explanation for all changes occurring to all matter, plant and animal life, which is responsible for evolution.

In my observations, I find there is not any voluntary activity that brings about any particle or change of matter. All matter, including plant and animal, has been force formed. A simple blade of grass has not voluntarily grown, but is force formed. There is not one animal, past or present, that was or is in existence voluntarily, but was or is force formed through a theory I have entitled, Perpetual Environmental Forces.

Scientific Fact—Barbara McClintock, Nobel Prize Winner

Barbara McClintock performed many experiments with corn. She found certain mutable genes appeared to transfer from cell to cell during the development of the corn kernel. She made a presentation to the scientific world exposing transferable elements. She proved that chromosomes and genes changed during the development of corn. It took her many years to gain this recognition. Now I theorize that these changes were due to Perpetual Environmental Forces acting upon the cells of corn. These forces act upon all matter which result in mutation and evolution. I base my theory on Sir Isaac Newton's theory of gravity, Charles Darwin's theory of evolution and the experiments with corn of Barbara McClintock which proved the chromosomes and genes (basic elements) changed during development. Although Barbara McClintock gave no explanation for these chromosome and gene changes, I explain these change phenomena with my theory of Perpetual Environmental Forces.

Definition of Forces and How Such Forces Result in Mutations As Explained By My Original Theory

All particles of matter are constantly being affected by multitudinous inundations towards matter in a push-pull, heating, cooling, blowing thrusting manner, perhaps physically changing matter, as a result of matter being acted upon by the sun's heat, cold temperatures (lack of heat), gravity, electromagnetism, out of space activity, cosmic rays, microscopic activity, and yet undiscovered forces, and also effected matter capable of affecting matter, which is a force upon matter. All of this activity is randomly and constantly occurring and in varying degrees with no deliberate pattern to all forms of matter.

Matter is always subject to action, reaction, and interaction, resulting from these random, constant changing activities or forces. All of this bombardment of activity I refer to as "Forces". I feel that all of this activity is a molding, manipulative happening acting and randomly forcing all particles of matter constantly. These forces act upon the genes in the reproductive area which can cause mutations and therefore make changes to further enable survival.

Therefore, each moment is different from each other moment. Once a moment passes, it is different from all past moments and will be different from all future moments. It would be impossible for an exact duplication of all forces and levels of forces acting upon matter, as matter itself is affected from momentary forces and this randomly affected changed matter also has an affect on matter. Although I used the word, momentary, I feel it is not a proper term, as the word momentary is a measurement of time, while I use the word to explain that rate of activity of force. Actually it is the activity or perpetual motion, that is constantly happening that appears to us as time. Time is the motion of forces perpetrating matter. Daylight is the sun's rays (force) shining or the rays moving toward the earth. As the sun's rays move towards earth with other forces along with the sun force, the earth's revolutions, gravitational pulls, electromagnetic forces, out of space activity, and undiscovered forces, these simultaneous forces are affecting the entire universe which includes our planet. With our naked eye, we see environmental changes taking place as day becomes night (absence of sunlight) and we call this time. Actually it is the movement of sunlight (a force) together with other forces acting upon our planet that we are observing.

Every particle of matter itself is a force. All matter occupying a place in space can act, interact and react to existing environmental forces. Its' composition determines its' state of being and the forces it can thrust. All matter is constantly affecting and being affected by perpetual environmental forces. When the sun (a force) heats the earth, each particle of matter receives the heat and in turn, this heated matter affects its' immediate environment by releasing some heat received from the sun. The immediate affect of this sun heat might have or not have a dramatic affect as to macroscopically change matter surrounding it, but it has some affect. We also know that the sun heat streaming down is a motion and therefore is not at a constant temperature resulting in continuously different action (force) on matter,

Example of Forces-Sir Isaac Newton-Law of Gravity

Another force is gravity. Gravity, as we know it on our planet, is constantly drawing all particles towards the Earth's center so that each particle of matter is constantly being affected by gravitational pulls. According to Sir Isaac Newton's "Law of Gravity", "gravity is a force of attraction by which terrestrial bodies tend to fall toward the center of the Earth". According to the lesson we learn in elementary school, Newton observed the apple falling off the tree. We are taught that this force, forcing the apple down towards the center of the Earth is gravity. It seems to me that if this same apple were to fall off the same tree and there would be at the same moment a large wind or even a wind with hurricane speed, the same apple would not immediately fall toward the center of the earth, but would come to rest at a number of different places far from the tree as a result of wind force and gravity force.

So that one can not think in terms of only this one force (gravity) acting upon matter (the apple) but that one must realize that there are many other forces simultaneously acting upon the apple and all other matter as well at every moment. The forces acting upon all matter can be so slight that macroscopically we see no change with the naked eye. But these forces are randomly and constantly acting upon all matter in varying degrees and can have dramatic affects as well as colossal catastrophic affects according to the forces and the composition of the matter existing.

Continental Drift-Platetechtonics- Alfred Wegener

The theory of continental drift (platetechtonics) proves that there is continuous movement, thereby continuous change within the interior of our Earth affecting the surface. But what is the driving force of this continental drift?

Alfred Wegener, father of the continental drift theory, said gravity alone could put continents in motion. I feel that gravity, (a perpetual environmental force) together with other perpetual environmental forces, as there is always continuous action, reaction and interaction, are acting upon the interior of our Earth causing movement or drift, earthquakes, volcanoes etc.

As a result of this continuous continental drift driven by perpetual environmental forces, the whole complexion of the Earth is forever changing. These changes can be so slight as to be hardly detectable and not disturbing or altering matter on Earth. But this continental drift can cause collision of continents which can cause catastrophic environmental conditions to life on Earth. As this continuous continental drift occurs which causes changes to the environment, such as land and ocean movement, resulting in climatic changes, all matter on Earth has to be able to survive these catastrophic events by means of their own force, acquired adaptable characteristics. The most adaptable to changes in the environment will survive leaving matter (plant and animal life) with less or no adaptable characteristics to die and so that the more adaptable to environment survive and can continue to reproduce.

If gravity together with other forces can move the interior of our Earth, then I believe these same forces must have an affect on all other matter on Earth. These forces act upon the genes in the reproductive area which can cause mutations and therefore make changes to further enable survival.

All Matter Not Identical

I have formulated my theory on the following existing phenomena. There is no existing matter that is exactly identical. Although different particles of matter are composed of the same elements, there is not one particle of matter composed of the same elements exactly duplicated. Each tree in the same species is different from one another and each leaf of the tree is different from each other and this is true of all matter. I believe that this is true due to the fact that every particle of matter is alone, occupying its own space, subject to random environmental forces. Every particle of matter is itself a force on other matter and is constantly in a state of being vulnerable to being affected or itself can affect other matter. The fact is that each individual particle occupying space has its own unique set of perpetual environmental forces inundating it, forcing it to be different. This thereby results in a situation where one of the same particles, being physically as close as possible to identically composed particles, are not identical. Under these circumstances there could be no identical randomly perpetual environmental force acting upon identically composed particle of matter. As a result, there could not be any identical particles of matter.

We know the air we breathe has a certain amount of weight, (one of its properties), bearing down on all matter. Obviously each place in air has a different weight bearing down on all matter. This is only one of many perpetual environmental forces acting upon each particle of matter. Therefore, many diverse perpetual environmental forces acting upon matter contribute to making each particle of matter physically unique that we can see with our naked eye and is different from one another. An example of this is snowflakes falling at the same time. One would suspect that such snowflakes might be identical to one another as they all are similarly composed of the same elements and falling at the same time in the same set of circumstances. But if we examine snowflakes under a microscope, they have certain similarities, but they are not physically identical. I believe the reason for snowflakes not being identical is that although each snowflake appears to be in the exact same set of circumstances, they are not. Each snowflake is independently force formed and acted upon independently and being affected by perpetual environmental forces due to the subtleties of occupying a different place in space. Examples of such subtleties are angle of the sun, or lack of sun, air weight bearing down, gravitational pulls, winds as well as many other forces. This is just a small example of a particular particle of matter, snowflakes, which is a simple composition. All matter and each individual particle is subject to the existing random perpetual environmental forces whether the forces inundating it appears to be identical or varied to any degree. This is due to the fact there is no identical spot, no matter how close each particle of matter is to each other. If you magnify this simple example to what really exists acting on each particle of matter, you can see that there is constant action, reaction and interaction resulting in no identical particles of matter. None of this force formed change is voluntary.

This does not mean that perpetual environmental forces are always changing and affecting each particle of matter. The outcome of any perpetual environmental force on matter is subject to the type and degree of force acting upon different types and degrees of matter. Simple formations of matter are most likely to be affected by environment frequently. As an example, the water vapor that forms into clouds and falls as rain, and rain, as the temperature freezes becomes snow, are changes to simple combinations of matter, resulting from environmental forces. There are different degrees of combinations of matter and how each is affected depends upon the construction of the matter and the forces acting upon it. With respect to human beings, which are extremely complicated structured forms of matter, we find that identical twins formed from one fertilized egg and developed until birth in the mother's womb, which we assume is an identical environment for each twin, macroscopically appear to be identical, but microscopically they are not identical, as each identical twin has a different set of fingerprints as well as other differences. Even though the original fertilized egg, which became two embryos, started simultaneously, the fact they have different fingerprints proves they are not absolutely identical and there has to be a reason for this occurrence. Since all identical twins have different fingerprints, although they are formed and composed from one fertilized egg, it would appear to me that at the time the identical twins' fingerprints are formed, perpetual environmental forces acting upon the moment of fingerprint formation, allow for different fingerprint formation to each twin by virtue of the fact that each place is different and affected differently by perpetual environmental forces, however slight they may be. These forces act upon the genes in the reproductive area which can cause mutations and therefore make changes to further enable survival.

"Big Bang" Breakaway and It's Results

What is the "Big Bang"? We know that in the universe there
are many objects. Occasionally an object falls to the earth from
outer-space such as an asteroid. Scientists believe before recorded
history, there was a grand explosion in the universe we call the "Big
Bang". This resulted in many particles disbursed. Our planet Earth
is one of those particles. I completely embrace the "Big Bang"
theory and start my theory from that point. Perpetual environmental
forces act upon all particles of matter regardless of its size or
construction. Therefore, these perpetual environmental forces have
been manipulating our Earth since the "Big Bang" breakaway.
I believe the "Big Bang" breakaway is the result of outer space
forces and whatever existed before which similarly was affected by
random forces, and have been acting upon all elements to form our
atmosphere and all matter including life on Earth progressing from
simple elements and cells to the constructed form of matter that
exists today.

Scientists have proven via scientific equipment that the Earth is
rotating and that there are several objects, (the sun, moon, planets,
stars, etc.) in our immediate outer space we call the Universe, and
that there are farer arenas in space called galaxies etc. I believe
the Earth is one particle or planet that broke away from this "Big
Bang" together with other particles which are all affected or forced
or manipulated by random perpetual environmental forces. These
forces are not deliberate, or purposeful, but move in a random and

involuntary way as a result of the motion and action, reaction and interaction of the forces manipulating matter.

I believe that there were and are certain conditions existing that are responsible for all matter on Earth starting with the simplest elements progressing or evolving to the most complicated matter existing today. These forces act upon the genes in the reproductive area which can cause mutations and therefore make changes to further enable survival.

"Big Bang" Activity on Earth

Scientists have also theorized that at the time of the "Big Bang" breakaway, the Earth must have been in a highly explosive gaseous condition, similar to an active volcano, which must have taken billion of years to cool down. During this cooling down period, these simple gases and elements were extremely vulnerable to change and must have been forced to change thereby combining, separating and interacting etc., caused by the forces that existed such as sun heat, or cooling due to the lack of sun heat, gravitational pulls, cosmic rays, forces within its' environment as well as outer space forces. Some forces are obvious (sun, gravitational pulls, wind and cold) as well as forces that we can detect with scientific equipment and forces coming toward Earth from outer space, as well as other forces that we can not presently detect, yet to be discovered. All the time, all matter is constantly in a state of being affected (forced) and thereby can change due to environmental forces. These forces can alter matter and it did and it still does, as this is an ongoing process. Nothing is ever the same. Each hour, each minute or each miniscule amount of time is different from the time before or the time after as all matter is subject to changing forces. Therefore there are constant changes and changing forces affecting all matter all the time. Some of these forces are so minute that they may not alter matter, but these forces do exist. These forces are in varying degrees and in time alter matter slightly or very dramatically in varying degrees and in turn alter matter slightly or very dramatically in varying degrees. Simple forms of matter are affected differently from more complicated structural forms of matter. If we just try to trace the affect of one force, the sun, in one twenty-four hour period on all matter, we can see it affects different matters differently, the affect it has on our air, the affect it has on water, the affect it has on plant life or on animal life. Different forms of matter are affected differently just from this one sun force and since there are many forces inundating matter all the time, no moment is ever the same as there are many forces individually and simultaneously acting upon all matter. As matter is formed, elements forced to combine to form all past, present and future gases, simple cells, plant life, animal life, etc., the combinations formed become more complicated and structured and can not as easily become changed by existing forces. Therefore there are many levels of matter being inundated by forces and reacting differently. These forces act upon the genes in the reproductive area which can cause mutations and therefore make changes to further enable survival.

Scientists tell us that our present atmosphere was formed by this interaction of being affected and forced by the environmental forces. I theorize that this formation is a continuously ongoing process. We know the gases and elements so combined to form the first simple cells for all life on Earth. Nothing was and is voluntary. As these simple cells formed, they could only survive if they, by means of force, acquired adaptable characteristics to survive in the environment. But we must not forget that these original cells were formed or forced by random environmental forces and these force-formed cells were continually affected by forces which is a continuously ongoing phenomenon.

Force Activity

Environmental forces appear to be the same as viewed by the naked eye. But the degree of their force is different and is always changing. Environmental forces can be so slight as we may not be aware of its existence or an environmental force can be so great as to have catastrophic affects. There are continuous forces all the time, pushing, pulling, molding, heating, cooling, building up, breaking down etc., which we are aware of as well as undetected forces. It is these forces that affect all life resulting from these forces and can allow life that formed to continue to exist or to change or to extinguish life, as we know it altogether. In addition, all matter is a force in itself. It acts upon its immediate environment. When sunlight forces itself on matter, this heated matter in turn causes an affect on its immediate environment, and together with other forces affecting this matter itself projects many forces on matter in its environment. Therefore there are many numerous perpetual environmental forces.

The mere fact that matter occupies space makes it a force itself. One grain of sand on a beach affects the other grains next to it. One grain of sand on top of another or underneath or along either side is in its own space being affected by perpetual environmental forces inundating it and is in turn thrusting a force around it, minute as it may be. Each individual particle of matter is alone with its adaptability acquired from forces, vulnerable to all existing forces. Just by observing the weather at any place, we see certain continuous change. We know, due to the earth rotating around the sun, (I theorize each particle of universal matter forced in place and subject to perpetual environmental forces) we can see with our naked eye, a most dramatic condition affecting to some degree all elements on Earth, the onset of day (sun-light) continuously changing until it becomes night (absence of sun-light).

Observations of Forces

There are great many conditions we experience affecting us. As
a result of this one forcing phenomena, changes in temperature,
wind, precipitation, all affecting us and most importantly,
constantly changing. When we go down to the seashore, we see
the ocean constantly in motion. We know there are gravitational
pulls (an environmental force) that keeps it down to Earth and we
can see with our own eyes and feel that the wind (a force) blows
the water. Therefore these environmental forces are affecting the
ocean. How many other forces are affecting the ocean for it to
be in a constantly moving state occupying space on Earth? We
know the sun due to its heat affects the ocean as well as the lack
of some heat. When we look at the ocean at the seashore, we see
the water constantly moving back and forth. It seems though, the
water coming towards the shore never comes back to the exact
spot as the movement before. It seems to me every water motion,
back and forth, is never the same as a result of environmental
forces never being the same as it affects the water. Even when
the atmospheric conditions appear to be exactly the same, the
water's motion is never exactly the same as it comes forward
immediately after going back to a different spot. Obviously,
there are subtle forces acting on the water as well as obvious
atmospheric conditions that we can see or feel due to winds or
rainfall. The fact that we know the water is always in constant
motion and always changing its degree of flow, back and forth,
makes it clear to me that there are always environmental forces

existing that are perpetrating water as well as all matter. Due to the fluid characteristic of water, we can see the immediate affect of environmental forces perpetrating it. These forces are also perpetrating all other matter at the same time, but all other matter reacts differently according to its force acquired composition and individual strength to remain the same, or to be altered slightly or dramatically or anywhere in-between. These environmental forces can alter matter at any time and may not alter some matter for hundreds or thousands or billions of years and might alter other matter frequently. But environmental forces are always affecting all matter all the time. These forces act upon the genes in the reproductive area which can cause mutations and therefore make changes to further enable survival.

When we look at the sky and see clouds, they never are identical. The water vapor of the clouds is a simple formation of matter and therefore is constantly being forced to change as a result of perpetual environmental forces as we observe them. Although there are similar cloud formations due to the similar atmospheric conditions forcing the clouds, they are not identical and they can not be because there are many different levels of forces acting upon the clouds at all times, as well as all other matter.

We know there are various climates existing in different parts of the world. We know that in desert areas there are certain plant and animal life that survives despite its low level of rain. In some areas there are rain forests and other areas frigid regions and again only certain plant and animal life survives. In order for life to exist in these extreme regions, this life must be adaptable to these extreme climates. It is the environment that determines what exists, as other life forms that are adaptable to other less extreme climates do not exist in these areas. Only matter (plant and animal life) that has been acted upon by random perpetual environmental forces, (which is an ongoing process) and acquires adaptable characteristics to survive in these extreme climates can, will and do survive. These forces act upon the genes in the reproductive area which can cause mutations and therefore make changes to further enable survival.

Adaptable Characteristics

Therefore, it is the random perpetual environmental forces that can alter all matter and as matter becomes altered, only matter that acquires adaptable characteristics to the environment can survive in extreme climates or any climate. It is not inevitable that some matter will be forced to acquire positive adaptable characteristics to survive in a certain environment, I feel it is just a matter of chance that these adaptable characteristics acquired by the action of perpetual environmental forces were formed and is certainly not voluntary. These forces act upon the genes in the reproductive area which can cause mutations and therefore make changes to further enable survival.

I feel you can examine any particle of matter, plant or animal and if it were endowed with more adaptable characteristics, it would be better able to survive in its environment. For example, if man were to have four eyes or four arms instead of two, and could fly and if birds had arms and fingers as well as wings, the life forms would be more adaptable to environment. It is only a matter of chance that the existing adaptable characteristics were acquired. But one must take into consideration all the formation and evolution resulting from random perpetual environmental forces that have taken place since the "Big Bang" till the present that allowed for all matter, as we know it existing presently or in the past, to have acquired the adaptable characteristics life possesses.

There is always a constant erosion to all forms of matter. There is always an end or death to all living creatures being inundated by forces. But at the same time, these forces are responsible for all past and present life and give rise to new life forms and matter on Earth due to forcing reproductive cells to change and recombining elements to form new matter. Plant life is a formation of simple elements that have combined due to many different forces inundating these simple elements and thus there is constant change as each change is again subject to environmental forces and to change, which is constantly happening. All change to matter has to be able to survive and reproduce in its environment or it will perish. I theorize that many similar organisms in a similar environment change similarly at the same time as they are affected by forces similarly being of the same composition and therefore, can reproduce with a like forced-changed organism. Naturally the newly forced-changed organism must be able to survive in environment. If there is a change and the change enables the organism to survive better in its environment, then the organism will survive and therefore reproduce with the newly acquired change caused by all matter being constantly vulnerable to change by perpetual environmental forces. If these perpetual forces change the organism so that it acquires characteristics that will hinder survival in its environment, the organism will not naturally reproduce. Therefore changes that enable for survival by making more adaptable matter (life etc.) in environment, result in the organism evolving and being endowed by better positive characteristics for survival. This does not mean the original organism will not survive and reproduce. It can survive if it has characteristics that enable it to survive as in its past, but the organism with the newly acquired more adaptable beneficial characteristics can survive and continue to be affected by the environment and can evolve to a higher form of life.

Beginning of Life Cells

All life started with simple cells, forced formed by changes due to perpetual environmental forces. All change takes place only in reproductive cells which are vulnerable to environmental forces as well as the rest of the organism, But the reproductive cells hold the formula and reproduces it in the next generation and so on. Since the reproductive cells, being vulnerable to all environmental forces as well as the rest of the organism's cells, but having the ability to reproduce, is the area where the environmental forces can and will change the organism ever so slightly or very dramatically due to the constant bombardment of environmental forces. The environmental forces have the power, by virtue of their nature, to alter the reproductive cells and so all forced-formed matter is always vulnerable to change. This all happens in an involuntary manner to the cells. When we look at certain plant life, ferns for example, we find a tremendous variety of a similar kind. It seems to me that the environmental forces that created plant life (ferns) were able to create a great many variety of ferns because the reproductive cells that gave rise to ferns are less complicated than higher forms of structured life and so existing environmental forces were and are able to more readily alter their reproductive cells and at the same time these forces allow for many varieties to appear and reproduce.

<u>Original Theory</u>

There are different kinds of mutations. An example of simple man-made mutations is the fruit nectarine which is an example of manipulating genetic cells of peaches and plums to form the nectarine. Different flowers can also be genetically manipulated. However there is involuntary manipulation as a result of mutations caused from perpetual environmental forces. Therefore, perpetual environmental forces cause mutations and are responsible for all life as it exists and will continue into the future.

Observations

When we look into the oceans we find a great variety of simple fish. Obviously the reproductive cells that form the simple fish due to environmental forces were able to give rise to a great many varieties as the forces allow for many varieties to be formed. In biology we are taught that a mutation is a dramatic change in an organism.

I believe, due to environmental forces, simple cells advanced to simple fish breathing with gills, and when environmental forces acted upon the reproductive cells of fish, a mutation appeared that evolved to lungs. This happening is involuntary. With lungs as an adaptable feature and also the force forming of arm and leg characteristics, these new creatures could leave the oceans and become land creatures with their new survival characteristics. This took a tremendous amount of time. Many different land animals having been further force formed, only the creatures which acquired positive adaptable skills were able to survive and continue to be vulnerable to further perpetual environmental forces. This is not a simple situation. All ocean and land creatures have to survive living among many other creatures and environment and only the fittest survive which have acquired positive characteristics to survive and reproduce. I theorize that most likely there always were and will be dramatic changes in organisms. But I believe the changes occur due to some dramatic change in environment. The environment determines the organism that will survive and since perpetual environmental forces can be dramatically catastrophic or any where in-between, then matter (plant life or animal life) can be forced to be changed dramatically or any where in-between as a result. These forces act upon the genes in the reproductive area which can cause mutations and therefore make changes to further enable survival.

Scientists tell us that all matter is composed of the same elements although they are not the same combinations. I theorize that all matter on Earth which are composed of the same elements, past, present and future, including all forms of plant and animal life, have evolved since the "Big-Bang" happening as a result of all matter being randomly manipulated by perpetual environmental forces. All evolutionary change can only be due to perpetual environmental forces, all of which is involuntary.

Although scientists do not know exactly how old our planet Earth is, they know it is approximately many billions of years old. I theorize during all this time perpetual environmental forces have been causing action, reactions and interactions upon all elements and all other matter resulting from these forces. In some instances, an element might change and might change back to its original form. Some elements might change slightly over long periods of time. Other elements might change dramatically in less time so that we may never know, as we check matter existing today, the changes it went through as the perpetual environmental forces that change matter were continually changing and the exact overall perpetual environmental forces that changed matter do not exist as they change from moment to moment. I believe that all life started as force formed elements, then forced to become simple single cells such as amoebae or paramecia and as environmental forces acted at random upon these cells, there were changes made.

The negative changes for survival died out and the positive more adaptable characteristics survived. There is a continuing forcing. Therefore, more adaptable characteristics can and were acquired so that higher forms of life evolved as a result of perpetual environmental forces, All matter that has been forced must acquire as a result of this force adaptable characteristics to survive in environment. These forces act upon the genes in the reproductive area which can cause mutations and therefore make changes to further enable survival.

The environment determines what kind of matter exists. Because the environment is in a constantly changing state all matter is vulnerable to constantly changing forces inundating it. We know there are many different species of plant and animal life. Each type of plant and animal life has different types of adaptable characteristics.

Other Observations

The word <u>instinct</u> is often used to describe an inborn pattern of activity or a natural or innate aptitude to a given biological species. I feel the word <u>adaptability</u>, (caused by the perpetual environmental forces) (not instinct) is a more appropriate word as it precisely describes any positive survival characteristics all species have acquired. And all species have acquired this adaptability as a result of perpetual environmental forces acting upon all species as well as all matter. The more adaptable characteristics that plant or animal life has acquired, being forced by perpetual environmental forces, this life will be better able to survive an environment. Naturally, characteristics that will hinder survival will naturally end in death and so life with adaptable characteristics survive. As life with adaptable characteristics survive, this life can become endowed with more adaptable characteristics as a result of perpetual environmental forces, resulting in higher forms of life. These forces act upon the genes in the reproductive area which can cause mutations and therefore make changes to further enable survival.

The acquiring of these adaptable characteristics appears as an orderly order of formed matter, which can be mathematically computed. But I believe this order is accidental, not deliberate and randomly forced into being. Scientists have been able to unravel many of the combinations and construction formed as a result of scientific probing and experiment. (This human intelligence itself is a force formed adaptable characteristic.)

Every particle of matter in order to exist has to have some rate of adaptability to environment. This adaptability to environment is a measurement of success of survival in environment. But as there always are changes in environment as a result of perpetual environmental forces, then success in survival is always questionable, as matter is always vulnerable to a continuously changing environment. Scientists in the past and present have found situations where there is an orderly, precise, pattern-like construction of formed life, both plant and animal. But as more experiments have been made and they delve deeper into this construction, they find a realm of less and less construction until there is a disorderly, non-precise, un-patterned like situation which scientists today call, "chaos". Now, to my way of thinking, "chaos" implies a negative situation where there is utter confusion or disorder or formless matter. Scientists have found an orderly, precise pattern-like construction of formed matter, and as they delved deeper, have found chaotic patterns. However, this orderly precise pattern-like construction and this deeper chaotic state are combinations of forced-formed matter. This so-called chaotic state is not a negative situation but a natural formation. From this chaotic state, being forced by perpetual environmental forces, an orderly, precise pattern-like construction can evolve and develop. In other words, this orderly or chaotic formation is a condition where matter is constructed with all kinds of acquired characteristics and therefore might be a simple element or simple forms of matter having acquired the adaptability to be in the state that it exists in relation to perpetual environmental forces inundating it. Once a change has taken place by being force formed, (provided the change is a positive change for survival), the matter is then orderly.

Adaptability

I feel there is a measurement, which I call adaptability, which describes the relationship between all elements, matter, plant and animal life, towards the perpetual environmental forces. This adaptability relationship precisely places the point of existence of each particle of matter. All particles of matter are at a stage of existence where they may be the simplest forms of matter with their adaptable to environment characteristics or any other stage of formation with their adaptable to environment characteristics and can acquire more adaptable to environment characteristics by being forced to change as a result of perpetual environmental forces and therefore could become a more constructed form and evolve into a higher formation. Therefore, if it were not for this orderly and so-called chaotic condition, there would be no evolution and no life as it exists today.

If our planet, Earth, was in an environment where there were no forces acting on Earth, then there could be no change on Earth from the original existing situation that existed at the time of the "Big Bang" or even before. The whole complete environment condition of forces acting upon the elements that existed on Earth and the perpetual environmental forces continuing to the present is responsible for all evolution of elements, matter and life on Earth as it exists today. Again, no change is ever voluntary.

Therefore, the perpetual environmental forces, because of their constant actions, determines what exists and all elements, matter, plant and animal life with their forced-formed adaptability are always vulnerable to stay the same, or be altered slightly or dramatically or become extinguished and exist in an adaptability relationship with these perpetual environmental forces.

As each particle of matter was formed (forced into being by random perpetual environmental forces) the forced-formed matter itself exerts a force on environment. In order to survive, all matter (plant and animal life) has to acquire adaptable characteristics. These adaptable characteristics result in forces upon other matter. When any life form ingests any particle of matter (air, water, food etc.) to survive, this is a force perpetrated upon other matter. As life was forced into being, it became a force itself in order to survive and as life forms evolved into more constructed forms, a greater force was perpetrated on other matter. As animal life became more constructed and their nervous system and brains became more constructed, the forces they perpetrated upon other matter were greater as they had acquired greater skills to survive. Naturally, the more adaptable animals survived and so higher forms survived and continued to acquire more adaptable characteristics. Human beings, at the present time, are the highest form of animal life and they can and do exert the most force on the environment.

Summation

I believe that all animals and fish that have brains can do some form of thinking and decision making. But human beings have evolved to the point of having language to communicate with one another which enables knowledge to be passed on to off-spring. Therefore, a great deal of knowledge is passed on to the next generation without the new offspring having to experience or discover all prior knowledge. This tremendous advantage continues to amass greater and greater amounts of knowledge. Originally man used all his efforts to obtain food. But as he was able to mass-produce food and other necessities resulting from the industrial revolutionary period, man has been able to allow his thinking abilities to go to a great many other areas, such as the sciences, medicine, arts, warfare, greater production of material goods for profit, etc. We know, as a result of exploration in the scientific field, man has created some of the greatest destructive forces, such as hydrogen bomb, chemically endangering the environment, etc. Therefore, there are multitudinous forces acting upon our planet, forces such sun, heat, outer space forces, cosmic rays, gravity, continental drift, microscopic activity, continuous actions and reactions of all matter and this matter affecting other matter. Now man, perpetrating forces on matter, plant and animal life), a forced-formed animal, is presently exerting a tremendous force on our planet. Man has the ability to exert forces which can be destructive to other forms of life as well as to himself. But man, by virtue of his forced-acquired adaptable characteristics of being endowed with a superior brain, can also use his skills, intelligence and knowledge to continue to delve into unexplained unexplored outer space and scientifically solve the mysteries of our planet for the betterment of all life on Earth. If we observe an infant we see that it is entirely dependent upon others for survival. The infant has the ability (acquired adaptable characteristics) to feel hunger and cry for food. As it is taken care of by its mother or others, it learns to depend upon and to be aware of these other, and naturally larger human beings, such as mother, father, or others that are physically close to it. As the infant develops, it becomes aware of these larger people and the direction it gets from these larger people and relies upon their instruction. There is a learning process and as the infant grows to become a small child and older child, it becomes programmed to being directed and to obey older and larger people. The infant grows up in an environment where it becomes accustomed to witnessing a supreme type person.

As the child grows older, it becomes aware of other supreme type persons with more authority than its parents, such as leaders of groups, during cave-men period, and other leaders as man's culture changed. During today's time, a supreme type person might be a person of religion, a mayor, a principal of a school, a president etc. In other words, human beings are conditioned and programmed from their first awareness to a higher, more authoritative situation existing in their environment. As man's brains developed and he was aware of his environment and could see some of the forces existing, such as storms, lightning, flood, drought, birth, death, sun, heat, cold, day light, darkness, etc., and not being able to understand these phenomena, but being programmed and conditioned to a supreme type being concept existing from infancy, it would seem natural for him to believe that all these unexplained conditions were the result of an unusually high supreme type all mighty being, higher in stature than any Earth being, and so religion came to be formed. Thus, through this religion concept, all unexplained phenomena could be unscientifically explained which gave man a great deal of comfort of having an easy fast answer to existing phenomena. On the other hand, teachings of religion also helped to civilize man,

Religion, as it evolved, also helped to direct man to create a culture and a society that was mutually beneficial to masses of people. As a result of this society and structure, man went forward in acquiring more collective knowledge and was able by means of experimentation to discover the reasons for many phenomena formerly unknown to man. But in today's time, although we are still programmed and conditioned to a supreme type person as a result of the fact that we start life as an infant, we should not continue to believe that existing phenomena that is now presently explained as result of proven scientific knowledge is a result of a supreme type religious act and to forge ahead in scientific endeavors to learn all the unknowns on Earth and the outer space arena.

I feel certain that this random perpetual forced type situation exists in the universe and galaxies beyond our planet. It remains to be discovered whether the identical forces we experience on our planet also exists on other planets, as there are many different forces existing yet undiscovered in outer space. As a result of the fact that there are many other different forces acting upon other planets, I believe it would be difficult or impossible to conceive of another planet with life forms as we experience on Earth. As the random perpetual environmental forces are responsible for the exact life on earth as we know it, to believe that there could be a planet with an exact duplication of the forces here on Earth would be highly unlikely. The forces on other planets determine what exists and it is possible to perceive some kind of construction. As man continues in his quest for knowledge, scientific exploration will reveal what exists beyond our planet and outer space.

ADDENDUM
WHY WE DREAM?

With respect to my original theory, PERPETUAL ENVIORNMENTAL FORCES, which cause mutations, dreams are an example of an adaptable characteristic discussed therein. Mutations occur as a result of my original theory, PERPETUAL ENVIORNMENTAL FORCES, which forces manipulate the reproductive cells of all living objects. Adaptable characteristics received from this forced mutation can and will enable the living item to further evolve. From simple elements to simple cells, being inundated by forces, mutations result in living items which now presently have many adaptable to environment characteristics. This action took billions of years.

Now, from these mutations, we know animals have many adaptable characteristics. Animals have eyes and can see. They have legs and can walk and run. In addition humans have hands with fingers that can do multitudinous tasks. Within the human body, there are countless adaptable characteristics that enable survival. One vital adaptable characteristic is the brain. In waking hours, the brain helps humans to survive in getting food, water, shelter etc. Now there is an involuntary period of time we call sleep. The sleep period is also an adaptable characteristic performing an important function for survival. What follows is my original theory of the DREAM period, an adaptable characteristic while we sleep and dream and how it helps animals to survive.

My theory explaining the function of dreaming by all living, thinking creatures. First of all, we know that all living creatures, in order to survive, must have physical characteristics that will enable it to survive in the existing environment in addition to nourishment, oxygen, proper temperature, etc. in order to continue to survive. Now let's start at the beginning of life—a newborn baby. It must have the physical characteristics to survive. In addition it has a brain, therefore, it is a thinking animal. One of the first reactions that a baby might have in its first environment would be temperature comfort or discomfort and once given nourishment, the comfort of food or discomfort of lack of food. The earliest infant dreams (I believe it dreams as it has a natural innate ability to dream) have to be related to the simple life around it. But as it develops and becomes more aware of its environment, its dreams expands in relation to the exposure around it. As the baby grows month by month it brain ingests more of the surroundings and it distinguishes between all pleasantries and unpleasantries. Physical and mental growth proceeds and there is a natural learning period by trial and error and a period of learning taught by mother (or others) to help the baby survive.—ex—not to touch hot objects, etc. As the baby grows to a child, it goes through periods of exasperation and frustration which will always continue as it learns to live with this for the rest of its life, as it grows and develops. As the baby grows into a child and its awareness of the structure of its culture and society becomes apparent to it, the child realizes its dependence upon parents and with this awareness there is a tremendous amount of frustration as it continues to grow and learns what is expected behaviorally towards parents and society. A child can perceive how it is naturally restricted physically and mentally by observing its elders. Each and every night the child dreams and deals with all the pleasant and unpleasant stimuli occurring during the waking hours. It is my feeling that in order for the child and all other thinking animals to survive the brain has this dream mechanism to be able to be completely free of the restrictions around it during its waking hours, and when it dreams it belongs solely to itself (in the dream state) and dreams of any situation of its brains choice. The brain directs the living creature and all physical functions are performed at the brain's direction; although some functions are involuntary such as heart pumping, but most physical functions are executed at the direction of the brain. (This is an oversimplified statement as the brain does not operate solely. It is a highly complicated organ affecting and affected by other organs—the nervous system; the chemical composition etc. of the body.) In essence the brain is the master of the creature and the physical activity is performed in accordance with the innate endowments of the creature. At the direction of the brain, the creature can only perform up to its physical potential. As humans our brains tell us we have the desire to physically fly as birds do, but we are not equipped physically to do so, so we can't and if we try we would become injured or die. As we are animals that communicate with one another, so we do not have to try everything that enters into our brain as we are taught self-preservation by our elders. The same situation applies to extended periods under water or in outer space, etc. In other words our brains can indulge in thoughts that are beyond our physical capability. There are two different beings—the thinking brain, that has untold potential horizons and the physical being which has enormous restrictions and cannot fulfill all the potential commands of the brain.

Undoubtedly this leads to a frustrated situation. If the physical being were to perform at the direction of the brain—flying, long underwater swimming and other unusual feats the human animal would die. But during the dream period the brain operates and can indulge in all areas whether or not these commands can be performed by the physical being during the waking hours. Dreaming is an involuntary function of the brain which is absolutely necessary in order for the brain to function normally to all kinds of stimuli during the waking hours. During the course of the waking period, living creatures are constantly in arenas where they are in constant contact physically and emotionally with their environment. By the time we reach adulthood, we become accustomed to our surroundings and learn to live with them and are not fully aware of all the constant changing settings we are in. If we were to become involved with all stimuli, we could not be able to function. It is the job of the dreaming mind to handle some of these stimuli so that fears, frustrations exasperations as well as pleasantries are explored by the dreaming brain, so that the waking brain does not have to handle these matters because in many instances these situations might result in death. It is an involuntary survival adaptable characteristic that we are equipped with such as heart pumping, breathing etc. I feel that during the dream period, the brain explores many situations relating to activities experienced during the waking hours such as physical frustrations and also personal relationships. As a baby the brain ingests and stores the stimuli in its environment and as it grows and develops so does the range of stimuli grows and the dream becomes more elaborate in relation to experience. There is an endless amount of areas that the brain can explore. This dream world is extremely important for the brain. In order for the awake brain to function for the survival of the individual, there has to be time for the brain to deal with and work out thoughts that it cannot be involved with during the awake survival period. If the awake brain were to be involved with all the thoughts it has and to explore them, there would be no time left to do what is necessary in order to survive. This is an adaptable characteristic that has evolved which enables survival during our waking hours yet the brain while we dream can deal with awake stimuli. During our awake hours we have to deal with our survival necessities such as—food, shelter, clothing. These necessities take up a good portion of our waking hours. But we are equipped with a dream period to peruse many different stimuli that might constrict us during our waking hours. Sometimes we remember our dreams and other times we don't. Most likely the brain must have a cut off signal as we do not remember all our dreams.

We have, because of Perpetual Environmental Force mutations, acquired an adaptable characteristic, during our sleeping dream hours, to deal with situations that might be harmful and can result in death if acted out during our waking hours. Therefore SLEEP activity performs a survival function.

Addendum Sexual Orgasm— Adaptable Characteristic

The following are natural conditions as apposed to artificial medical procedures. We know that within the female body there is an egg which is microscopic. This egg can expel in approximately thirty (30) days. Female puberty most often occurs any time between the ages of ten (10) or sixteen (16). In the male body there is a scrotum at the genital area which can ejaculate sperm. These sperms are also microscopic.

It seems to me the egg within the female body might be similar to the original force formed egg that evolved after the "Big Bang", the scientific theory, happening. This egg is awaiting a sperm. If no sperm penetrates the egg, it is expelled. The sperm is a force trying to get to an egg to fertilize it. If the sperm reaches the egg within the female body and penetrates it, fertilization begins. This fertilized egg continues to develop until it is ready to be born. This is called pregnancy. The whole action after fertilization occurs show an evolution from a fertilized egg to a human being. This shows how humans were developed.

It is a known scientific fact the fertilized egg progresses to a fetus that at some time has gills, as fish have, and a tail, that animals have. There are other physical features, the fertilized egg has, and that it does away with. It continues to grow and develop with adaptable characteristics.

During the evolutionary process in animals as well as human beings, there came about a female with eggs within her body and a male with sperm cells within his body in the scrotum area.

In order to have reproduction, we know the male must ejaculate sperm into the female. The female has a vagina for entry. The male has a penis for this purpose for entry into the vagina. We call this sexual intercourse. A sexual orgasm with its pleasurable moments can result in sexual intercourse. Humans are emotional beings. Therefore relationships come in to play. The desire for one another (love) is a beautiful emotion which usually brings the partners together. The adaptable characteristic the male has is his strong desire to have an orgasm. The adaptable characteristic the female has an ability to be desired to attract a male for the sperm. As a result, physically, she to has strong desire to have an orgasm. Perhaps without this strong sexual orgasm pleasure, humans might not continue to propagate the species. Now in order for this to happen there was a PERPETUAL ENVIRONMENTAL FORCE mutation to bring about this situation. My original theory PERPETUAL ENVIRONMENTAL FORCES is explained in the beginning of this book. A sexual orgasm, with its pleasurable attributes, can result in sexual intercourse.

With regards to my original theory, PERPETUAL ENVIRONMENTAL FORCES, which cause mutations, sexual orgasm is an example of an adaptable characteristic which enables procreation and survival of the species. There is no activity that compares to the ultimate thrill of an orgasm.

One may partake in eating dark chocolate and other rich deserts such as ice cream or cake but the pleasure from these sweets, does not compare to an orgasm. Some enjoy foods such as filet mignon or porterhouse steaks. Lobster and salmon are also favorites. Others might enjoy Italian, Chinese or Thai foods. Do these great tastes compare to an exciting orgasm? Jewelry, diamonds, clothing and pursuit of wealth are desired by people. These material items do not compare to an orgasm. Some people enjoy and live in palatial homes. How thrilling is this compared to an exciting orgasm. Others take expensive trips to all parts of the world. The above are examples of activities people indulge in and sometimes they are orgasm substitutes. Now we find people in poverty can also enjoy orgasms by virtue of their physical adaptable characteristics. You will also find people incarcerated also have this adaptable characteristic and so they are not limited. As a result of PERPETUAL ENVIRONMENTAL FORCES there is an adaptable characteristic, an orgasm which enables procreation. An orgasm is the ultimate in sexual pleasure which can result in procreation. PERPETUAL ENVIRONMENTAL FORCES as it causes mutations developed an orgasm state which enables survival of the species. Orgasm is an adaptable characteristic. The proof of this is our large population due to the desire of this ultimate pleasurable orgasm activity. The orgasm thrilling pleasure is an adaptable characteristic resulting from mutations caused by PERPETUAL ENVIRONMENTAL FORCES.

About the Author

Sylvia Goldstein was born, raised and educated in the City of New York. She attended New York Institute of Applied Arts and Science and Queens Community College. She is a philosopher, writer, poet, essayist, artist, stone sculptress, photographer, pianist and paralegal consultant. Organized Current Events Club and President of same for 12 years. Resided with her husband, David Goldstein, Esq., attorney at law and family in Great Neck, Long Island, New York. They presently reside in Boynton Beach, Florida and Atlantic City, New Jersey.

dedicated to David Goldstein

ORIGINAL STONE SCULPTURE
By Author SYLVIA GOLDSTEIN

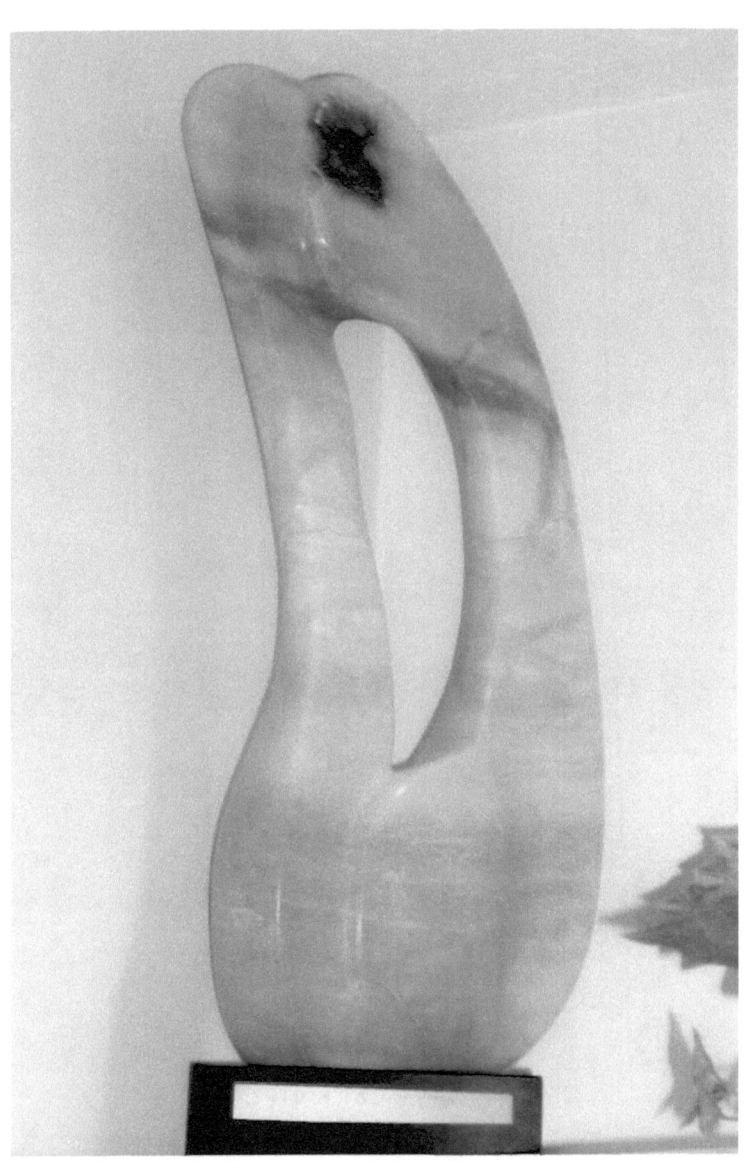

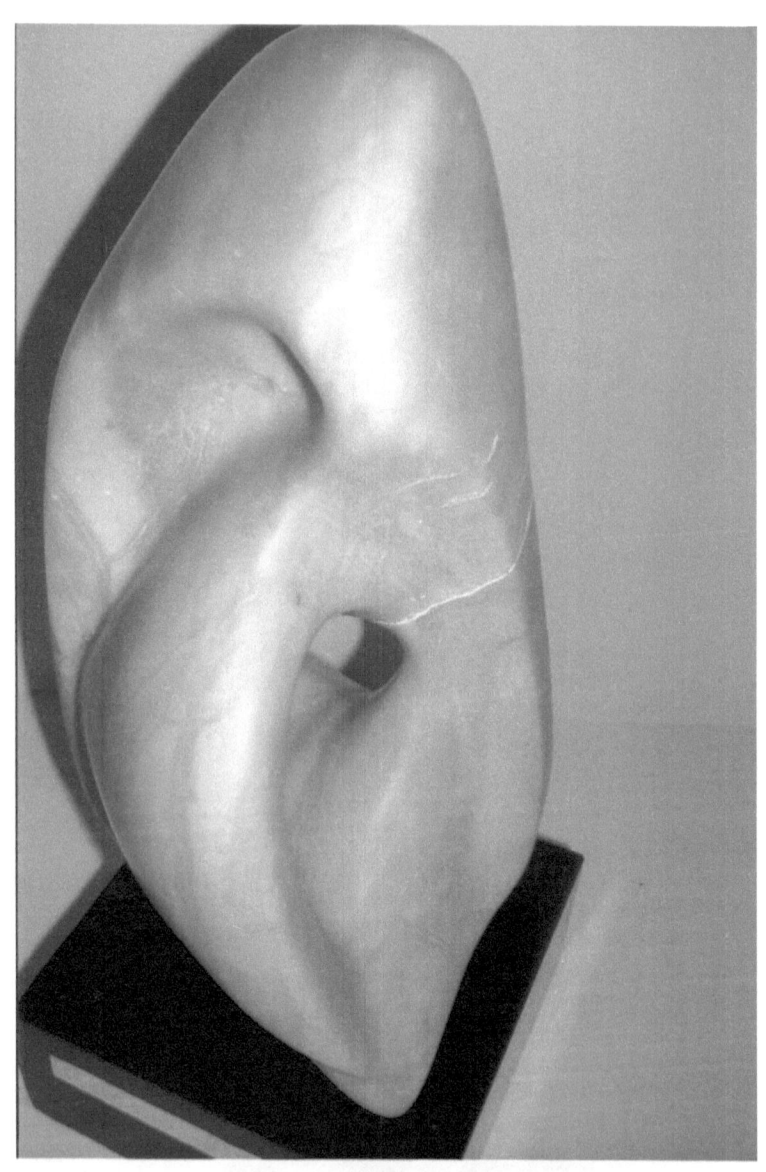

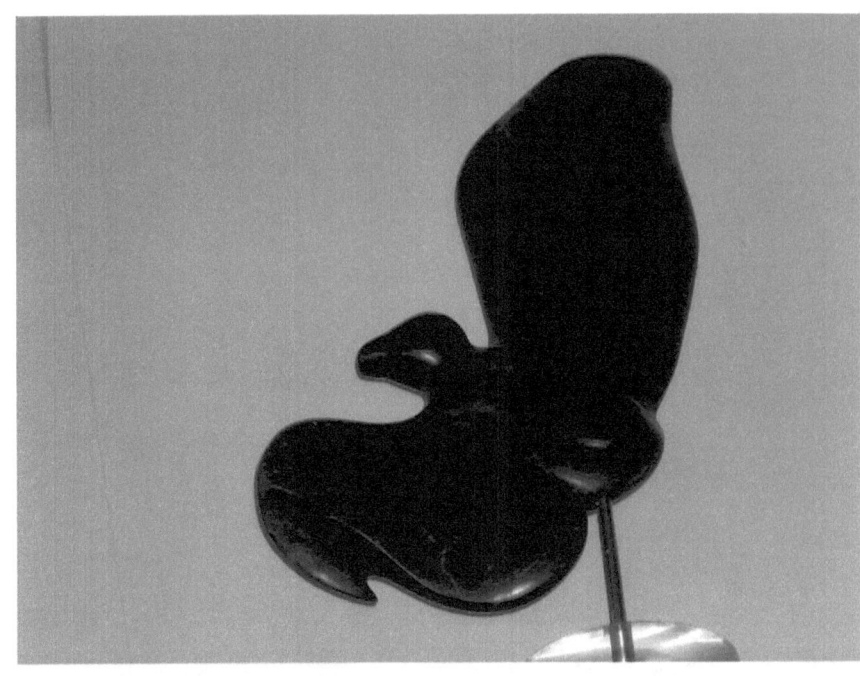

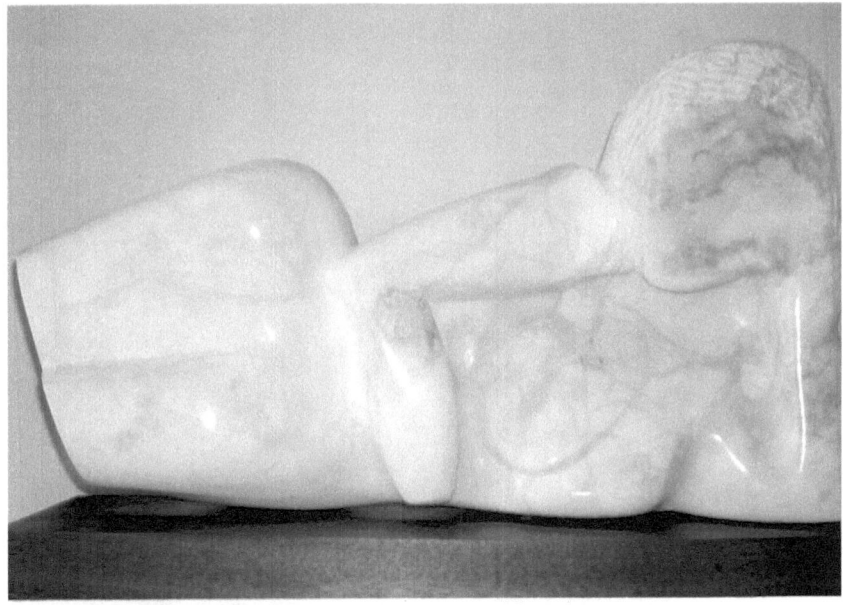

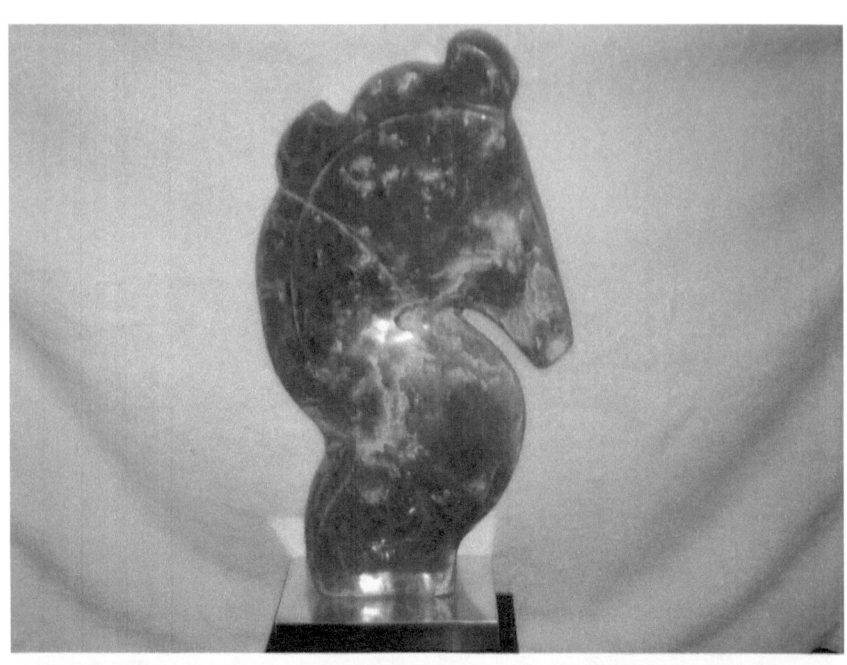

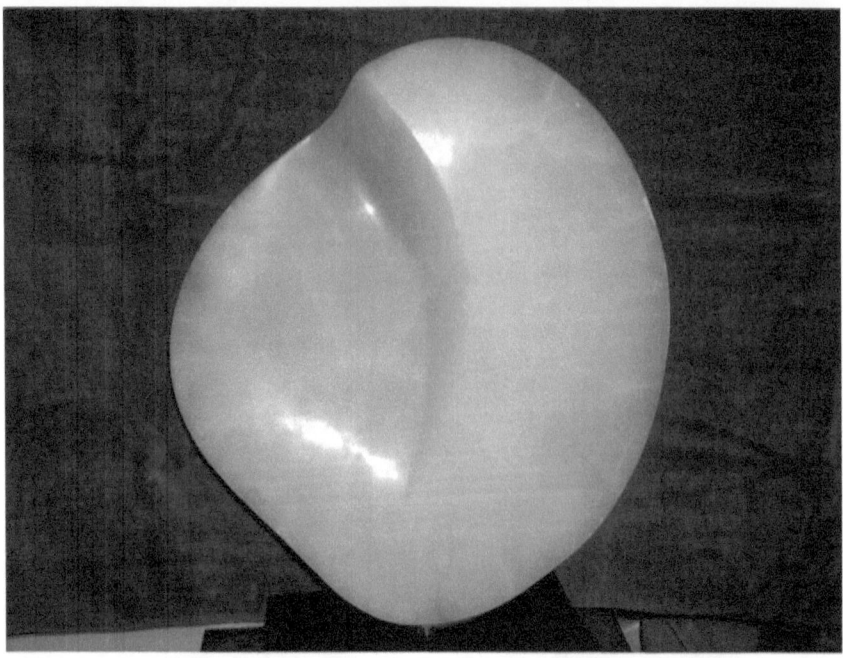

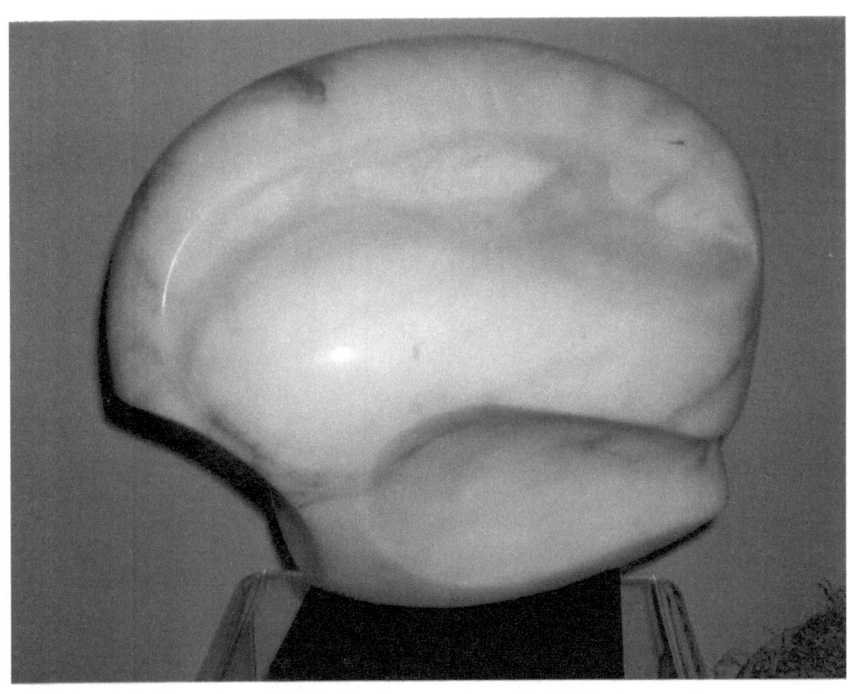

ORIGINAL ART MEDIA
By Author SYLVIA GOLDSTEIN

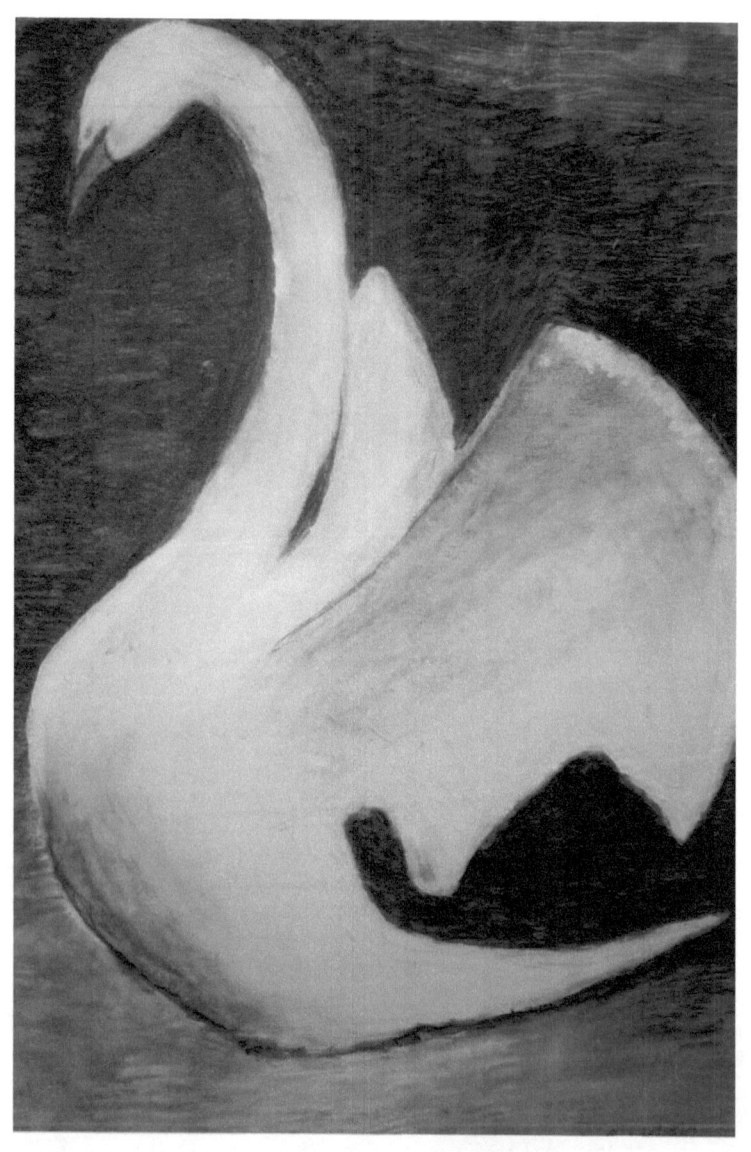

www.ingramcontent.com/pod-product-compliance
Lightning Source LLC
Chambersburg PA
CBHW021909170526
45157CB00005B/2023